Library of
Davidson College

What the experts say about *Track and Field Dynamics*:

"*Track and Field Dynamics* is a great book—necessary if one is to understand the mechanics of track and field. A must for all competitors."

> —Ralph Boston
> Olympic long jump champion

"Track and field in the United States has needed a new approach. This book fills that need in many ways. It is extremely practical and is invaluable to both athlete and coach."

> —Ron Morris
> Olympic pole vault silver medalist
> Track Coach, California State
> College at Los Angeles

"*Track and Field Dynamics* deals with the basic laws of physics for track and field events in simple, understandable terms for the coach and athlete. An outstanding contribution to the sport of track and field."

> —Payton Jordan
> 1968 U.S. Olympic Coach
> Track Coach, Stanford University

"*Track and Field Dynamics* is a great book . . . Tom Ecker has made it possible for track coaches (and perhaps most important of all, track athletes) to easily understand the mechanics appropriate to movement in track and field. I am certain that this text will be a stimulus to help athletes reach their potential."

> —Stan Wright
> Asst. 1968 U.S. Olympic Coach
> Track Coach, Sacramento State College

"Few scientists who understand body dynamics have the time, the interest or the know-how to apply these scientific principles to the sport of track and field. Most of our authorities in track and field don't have the time, interest or know-how to be able to write or teach in the area of dynamics. Tom Ecker has bridged this gap in his new book, *Track and Field Dynamics*."

> —Clarence Robison
> Track Coach, Brigham Young University

"Tom Ecker is one of the most analytical writers of track and field in the United States. In *Track and Field Dynamics,* he has taken the difficult and made it easy. Coaches and athletes will gain new insight into track and field mechanics, while the manner in which the book is written will enable the sincere fan to gain increased appreciation for the performance of track and field skills."

—Harry Groves
Track Coach, Penn State University

"A long over due attack on many popular 'coaching misconceptions,' this book presents, in an interesting and terse manner, a much needed 'Advanced Course' on the mechanics of track and field events. It is a book every serious minded coach should read and apply."

—Bill Perrin
Asst. Track Coach, U. of Wisconsin

"Tom Ecker's *Dynamics* is a unique, long needed and important contribution to American track and field literature. It calls the coach's attention to the true fundamentals of the sport and, most importantly, it helps him to reject unsound theory and faulty technique."

—Bill Huyck
Track Coach, Carleton College

"Too many coaches and athletes attempt to 'copy the champion' without taking into consideration the reasons behind aspects of the champion's form. Knowledge of the mechanical principles and the physics involved in reaching for excellence in track and field events is an absolute must. Tom Ecker is a leader in imparting this knowledge to American coaches and athletes."

—Berny Wagner
Track Coach, Oregon State University

"A much needed area, the physics of athletics, comes to life in Tom Ecker's newest and best book."

—Jim Terrill
Track Coach, Amherst College

"Required reading for the coach who really wants to help athletes reach their potentials."

—Chuck Rohe
Track Coach, University of Tennessee

TRACK AND FIELD DYNAMICS

Tom Ecker

TAFNEWS PRESS

Book Division of
Track & Field News

First published in 1971 by Tafnews Press,
Book Division of Track & Field News,
Box 296, Los Altos, California 94022 U.S.A.

Copyright © 1971 by Tom Ecker
All rights reserved.

Library of Congress Catalog Card Number: 76-143565
Standard Book Number: 911520-22-8

Printed in the United States of America

Things are not always what they seem.

— Phaedrus (circa A.D. 8)

Tom Ecker (left), shown above with his long-time associate Fred Wilt, has had a varied and colorful career in track and field. As a high school athlete in Iowa, he won eight individual state championships, setting state records in seven of them. He won championships in the Big Ten and at the Drake and Kansas Relays during his collegiate days at the University of Iowa.

Ecker coached at the high school level for three years before beginning a four-year stint as track coach at Western Kentucky University, where he was named Ohio Valley Conference coach-of-the-year all four years.

The author of six sports books and more than 70 technical magazine articles, Ecker has traveled widely throughout North America and Europe, lecturing on various aspects of sport. His humorous and informative talks have taken him before several state coaching associations, as well as two NCAA national clinics, the AAHPER national convention and meetings of international sports delegations in London, Mexico City, Stockholm, and Munich.

In 1966 Ecker was hired by the Swedish Amateur Athletic Association as national track and field coach of Sweden. After a year in Sweden, during which time the Swedish National Team completed its first undefeated season in 18 years, Ecker returned to the USA to become Coordinator of Athletics in the Cedar Rapids, Iowa, school system.

Foreword

Indifference is a quality of which Tom Ecker is incapable. His personal competitive and coaching careers already sparkled with success when he became fascinated with the application of classical Newtonian laws of motion to track and field skills as an avenue to improved performances.

His enthusiastic pursuit of knowledge in this area led him to attend many lectures delivered on the subject by Britain's world renowned former Chief National Coach, Geoffrey H.G. Dyson, in the USA and in Canada, and to study in the classes conducted by British physicist Bernard Hopper at the Royal Canadian Legion National Coaching School during a number of summers.

In an honest search for truth, Ecker has availed himself of many of the volumes written on Newtonian mechanics, biomechanics and kinesiology. Not the least of these has been "Mechanics of Athletics" by Dyson.

It is interesting to note that no American author has previously produced a book dealing exclusively with the application of Newtonian laws of motion to track and field skills.

One reason Americans often neglect to apply mechanical laws in practical and theoretical coaching is their apparent lack of basic knowledge of these laws. Much of the literature currently available on the subject of biomechanics is written with such unintelligible nomenclature that coaches are often unable to understand it. Tom Ecker has come to grips with the problem of comprehension in this volume. He has cut through the "red tape" of technical jargon in many respects, to bring to the track and field coach a practical approach to the application of mechanical laws to his sport.

Readers with prior knowledge of physics will notice that he elects to discard certain words of mechanical jargon in favor of terminology more relevant to the coach's vocabulary.

Unfortunately, professional jealousies sometimes exist in sports literature. Though seldom articulated, a subliminal impression may sometimes be transmitted that if the self-designated "authority" did not "do it," "say it," or "write it," then it should not have been said, done or written.

Ecker, however, makes no claim as an authority in physics, mathematics or science, and readily extends full credit where due to the giants in scientific and sports literature who have preceded him. He has, by virtue of diligent study and an inquiring mind, sought to bring order from what for many may have been chaos. For this I salute him.

— Fred Wilt
Editor, *Track Technique*

Preface

The evolution of sports techniques through the ages has followed a rather inverse, paradoxical pattern. In theory, coaches devise or discover athletic techniques or technique changes, and then impart them to their athletes. But the opposite has always been true—from the time the first running strides were taken, back in the days of pre-history, to the development of the Fosbury Flop.

An athlete (or a number of athletes over the years), through trial and error, develops (or improves) a particular style. Occasionally, if a change is particularly unusual, the athlete is given credit for his innovation (O'Brien's shot putting style, Dooley's pole bending, Fosbury's Flop, etc.).

In many cases, a technique exists because all motion must obey the laws of physics. Error is impossible, and future coaching of those particular techniques, except for polishing, is not really necessary. The laws of physics are as irrefutable for a hurdler whose left arm and right leg are forward during hurdle clearance as they are for a rabbit being dropped from an upside-down position. The hurdler's left arm goes back as the trailing leg comes forward; the rabbit turns over in the air. No one had to coach either movement. Because of the principles of dynamics—the branch of physics which deals with the action of force on bodies—both movements are inevitable.

This book does not cover the entire realm of dynamics. Some areas that are important to the physicist are being left out because they are not important to the track and field coach. Also, for simplicity, some of the physics terminology has been altered. Included here are only the laws and principles that relate directly to track and field coaching, or that are necessary as background information in learning those laws and principles, in simplified language.

Most of the ideas in this book have been gleaned from (or inspired by) the works of Galileo, Sir Isaac Newton, Dr. C. H. McCloy, Geoffrey Dyson, Bernard Hopper and Fred Wilt. My role has been to apply, simplify and clarify.

Some of the interpretation and application has come from my brother, Dr. Richard E. Ecker, who is himself both an outstanding physicist and a track and field enthusiast.

The study of dynamics is a whole new world—an exciting world—with much still to be discovered and explored by scientific-minded coaches who are not afraid to probe into the unknown. The possibilities are boundless, limited only by the coach's own imagination.

— Tom Ecker

Contents

	Foreword by Fred Wilt	7
	Preface	9
1.	Introduction	15
2.	Motion	18
	Linear motion	
	Rotary motion	
	Acceleration and Deceleration	
	Forward Lean	
	Acceleration and forward lean	
	Air resistance	
	Component Velocities	
3.	Center of Mass	28
4.	Curves of Flight	32
	Parabolic Curve	
	Optimum Angles	
	Aerodynamic Forces	
5.	Axes	39
	Axis of Momentum	
	Axis of Movement	
	Nutation	
6.	Momentum	46

7.	Inertia	49
8.	Conservation of Rotary Momentum	51
9.	Turns from the Ground	61
	Eccentric thrust	
	Checking linear motion	
	Transference	
10.	Turns in the Air	72
11.	Inertial Axes	80
12.	Falling Bodies	86
13.	Secondary Axes	91
14.	Action-Reaction on the Ground	96
	Impulse	
15.	Centrifugal Force	103
	Glossary	105
	Bibliography	108
	Index	109

Acknowledgements

My thanks to Vicki McAllister for her typing, to Betty Spaight for her artwork, to Jack Griffith for his many fine photos, to Fred Wilt for his guidance, and to Judy Ecker for her criticisms (some of them constructive).

CHAPTER 1

Introduction

Coaching is an art; no one will dispute that. Nor will anyone dispute the fact that there are many clever and colorful "artists" in the U.S. coaching community.

The art of track and field coaching embodies three sciences—physiology, psychology and dynamics. American track and field coaches have become quite knowledgeable in the area of physiology, and have long been known as the best practical psychologists in the world. It is the area of dynamics, the branch of physics that deals with the action of force on bodies, that has somehow been neglected in the American track and field coaching scene.

How many coaches still believe that a straddle high jumper can create rotation over the bar and help to clear his trail leg by forcing his leading arm down toward the pit or back under the crossbar? Or that it takes strength to keep the legs up just before landing in the long jump pit?

That a shot putter can determine the angle of release by the angle of his "strike" against the shot?

That sprint lean can and should be taught?

That a high hurdler should keep his head up and look ahead as he clears each barrier?

That long jumpers should always strive for height,

attempting the ultimate of 45 degrees?

That sprinting speed is an inborn trait and cannot be appreciably improved?

That long jumpers should reach forward with their arms just before landing?

That a high jumper should jump without a shoe on his leading foot?

That it is possible for a sprinter to run a 100 or 220 and continue accelerating all the way to the finish line?

That a strong following wind actually "blows" the runners along?

That gaining height is as important in the long jump as take-off speed?

That the hitch-kick is not really important, except for rhythm and balance?

That high hurdlers have a greater balance problem than low or intermediate hurdlers?

I mention these particular points because I believed all of them to be true when I began coaching. I believed them because they *seemed* to be true. Now I know that all of them are false, simply because they all go against the laws of dynamics—laws that are scientifically beyond argument. All motion in sports must follow the same dynamical laws as everything else on earth.

Yet we still have "non-believers" at every level of coaching, even though dynamics is the most exact of the three scientific disciplines for coaches. Because "It doesn't seem right" or "That's not what *my* coach told *me!*," there are coaches who cling to parochial beliefs and continue to coach in a way that can only be considered charlatanistic, at best.

Dennis Horne, the trampoline expert, has written, "The human senses have been known to play many strange tricks on us from time to time. By relying upon those alone we may find that we imagine and believe that a certain action is producing a certain movement, when scientifically this can be proved impossible."

The laws are there. They must be obeyed by human movement, either with or without the knowledge of the coach.

The importance of dynamics was best presented by Geoffrey Dyson, the brilliant coach, when he wrote that the purpose of a sound knowledge of dynamics is "to distinguish between important and unimportant, correct and incorrect, cause and effect, possible and impossible."

And, after all, that's what coaching is all about.

CHAPTER 2

Motion

There are two types of motion—*linear* and *rotary*. Both come into play in track and field.

Linear motion is motion along a generally straight line, such as the path of the sprinter in a dash race. All parts of the person (or object) move the same distance, in the same direction.

Rotary motion is turning or rotating motion in a circle (or arc) around an axis, such as the rolling movement of the straddle high jumper as he clears the crossbar or the spinning of a discus in flight. All of the mass outside the axis (which runs through the jumper's body or through the center of the discus) is in motion around the axis, while the axis itself remains in a fixed position, relative to the rotation around it.

A car moving along a highway is an example of linear motion, traveling from one point to another along a straight line, but its wheels, which rotate around axles, demonstrate rotary motion.

In track and field, every event requires both linear and rotary movements. For example, the linear motion of a runner is coupled with the rotary motion of his swinging arms (around an axis through his shoulders) and his driving legs (around an axis through his hips). (See Figure 1.)

Figure 1. The runner's motion over the ground is linear, but the motion of his arms and legs is rotary, around axes through his shoulders and hips.

Jumpers attempt to achieve great height or distance (depending upon the event) in a linear direction and are aided by various rotary movements. Weight men use rotational movements (rotary motion) to accelerate and release their implements in a linear direction.

ACCELERATION AND DECELERATION

Acceleration is rate of change of velocity. In sports, velocities are seldom constant, and so we must think in terms of gaining speed and losing speed. An increase in velocity is called *positive acceleration*. In the field of physics, a decrease is called negative acceleration, but it is usually called *deceleration* in the coaching world.

As a sprinter leaves the blocks, his acceleration is very great at first. But as the ground begins to move faster and faster under the sprinter's feet, (and as wind resistance increases with every accelerating stride), he must reach a point when he cannot move his legs fast enough or keep his feet on the ground long enough to continue accelerating. At that point (usually 60-85 yards from the start), the sprinter's legs are moving so fast that the efficiency of the muscular contractions is only one-fourth of what it was at the beginning of the race.

The sprinter may maintain his top speed briefly, but because top-speed muscle contractions are possible for only a few strides (limited by muscular endurance), he begins a gradual, unnoticed deceleration, no matter how much he tries not to, all the way to the finish line. It may appear to the spectators, and feel to the sprinters, that they continue to accelerate to the finish line, but it simply cannot and does not happen.

When the sprinter reaches the point of maximum acceleration in his race, his rate of deceleration from that point on is determined by the level of his muscular endurance. Everything else being equal, the man who is better conditioned "slows down the least" and wins the race.

There are coaches who claim that certain sprinters have the amazing ability to accelerate at the finish of a 100- or 220-yard race. If these sprinters are running their races all out, without loafing, then acceleration at the end of the race is impossible, according to the laws of physics. Once the legs are moving at their fastest possible frequency, obviously it is not physically possible to increase that frequency, or to increase stride length without reducing that frequency. Therefore, the only way to accelerate at the end of a sprint race is to deliberately *decelerate* at some point in the race or deliberately *not accelerate to maximum* from the beginning, either of which will contribute to running a slower total time for the race.

FORWARD LEAN

Forward lean in running is not something that can be coached or learned. It is the direct result of the positive acceleration of the runner and, to a lesser degree, air resistance.

Thus, it is never necessary for a coach to give his sprinters such advice as "Stay low out of the blocks," "Lean forward at the halfway point," etc. Sprinters may assume a more bent-over position, in an effort to increase forward lean, but actual lean—the line between foot contact and center of mass—remains the same. Except for the slight natural, uncontrollable lean against air resistance, forward lean can be increased by the athlete only through positive acceleration, and that is possible only for a short period of time.

Acceleration and forward lean. Disregarding air resistance for the time being, there is an exact correlation between acceleration and forward lean. The greater the acceleration, the greater the lean. (Acceleration is the cause and lean is the effect; never the opposite.)

Forward lean in sprinting is greatest early in the race, when the sprinter is accelerating rapidly. (See Figure 2.) While acceleration decreases to zero (60-85 yards from the

Figure 2. The sprinter accelerates rapidly from the blocks and has a correspondingly great angle of forward lean. As acceleration is reduced, the angle decreases.

start) and deceleration begins, body lean reduces automatically.

In the closing stages of a 220-yard race, where the rate of deceleration is particularly high, it is even possible for the sprinter to have a slight *backward* lean. (See Figure 3.)

When using a photograph to determine angle of body lean, always select a "mid-stride" photo, with the knees and arms in the same approximate position. In any other part of the running stride, it is impossible to determine angle of lean visually.

Air resistance. The other factor contributing to forward lean is the resistance of the air against the sprinter. As speed increases, so does air resistance, and, to counteract the tendency to fall backwards, forward body lean must increase

Figure 3. A backward lean in the closing stages of a 220-yard race is the result of deceleration. Even though the runner faces air resistance of approximately 20 m.p.h., the forward lean into the air resistance is more than offset by the effects of deceleration. (This photo shows the closing stages of a 20.3 220-yard race.)

slightly. When the sprinter faces a headwind, his lean must be more acute. When he has a trailing wind, his posture is more upright.

In theory, if a sprinter who has reached maximum acceleration is running at 30 feet per second and has a trailing wind of 30 feet per second, there would be no forward lean at all, no matter how much the sprinter might try to lean forward. If the trailing wind were 31 feet per second, he would actually have a *backward* lean.

Physicists have computed that approximately 18 per cent of a sprinter's total effort is expended on pushing air aside. When there is less air to push, as in wind-aided races or in races at altitude, a faster speed can be achieved with the same energy expenditure.

In wind-aided races, it is interesting to note that the wind does not blow the runner along, as is often believed; it merely gives him less wind resistance to face. A sprinter running at 20 m.p.h. with a 5 m.p.h. trailing wind has to face only 15 m.p.h. of air resistance, instead of 20.

COMPONENT VELOCITIES

When an athlete leaves the ground and is free in the air (such as in hurdling, high jumping, long jumping, etc.), two completely separate velocities have been imparted to his body—one horizontal and one vertical. Separately, they would provide either straight-forward speed or straight-up lift, but the two component velocities together, whether they have been imparted separately (as in long jumping) or simultaneously (as in sprint starting), provide a resultant velocity and an angle for leaving the ground. (See Figure 4.)

Figure 4. The hurdler is not conscious of it, but he imparts two separate velocities as he takes each hurdle—one forward and one upward—which determine at what velocity and angle he will leave the ground.

The same is true of the release of throwing implements. In the shot put, for example, the *shift* across the circle provides the horizontal component; the *lift* of the putter provides the vertical component. The resultant velocity, which is a combination of shift and lift, determines the release speed and the exact angle of release, no matter how much the putter might want to raise or lower the angle as he "strikes" the shot. (See Figure 5.)

Figure 5. The shot putter imparts horizontal velocity first, then vertical velocity. Together they determine at what velocity and angle the shot will leave his hand.

If the putter is fast across the circle, but does not lift with as much vertical velocity as the horizontal velocity he has achieved, the shot must be released at a low angle.

If the putter is slow across the circle, but lifts with more vertical velocity than the horizontal velocity he has achieved (which rarely happens), the shot is released at a high angle.

Of course, the ultimate distance of the put is determined by two factors—the angle of release and the velocity of the shot at the time of release. Therefore, to bring about an improvement in distance, either the shift or the lift velocity may be improved for an increase in resultant velocity, but in order to maintain the proper release angle, neither velocity can be significantly greater than the other.

It is important to note that the application of force to the shot must always be directed through its center of mass in order for the resultant velocity to be at its maximum. If the force is applied alongside the center of mass, the shot will take on rotary motion, its velocity will be diminished, and it will not go as far.

CHAPTER 3

Center Of Mass

The center of mass (often called center of gravity) of an object is the point where all of its mass is concentrated. The weight of the mass on any one side of the center of mass must be equal to the weight on the opposite side.

The center of mass of a 16-pound shot is presumed to be at its very center, with eight pounds of weight on any one side of that center and eight pounds on the opposite side. A doughnut's center of mass is in its center, but it is also in space, showing that the center of mass of an object does not necessarily have to be *within* the object's mass.

In the human body, the center of mass is not a fixed material point located in a specific part of the body. As body position changes, the center of mass changes its position within the body.

In a standing position (Figure 6a), a person's center of mass is approximately one inch below the navel. (This depends, of course, on his weight, body shape, etc.) By raising both arms (Figure 6b), his center of mass is raised about five inches. Bringing one arm down (Figure 6c) lowers the center of mass one and one-quarter inches and moves it to the side. Figure 6d shows that the body's center of mass does not necessarily have to be within the body.

Figure 6. The center of mass of the human body is the point where all the body's mass is concentrated. As body position changes, so does the center of mass.

There are a number of track and field events in which the body's center of mass is outside the body at some time during the execution of the event. In long jumping, for example, the center of mass is within the body during most of the flight to the pit, but shifts forward, just outside the body, as the jumper assumes his landing position. (See Figure 7.)

Figure 7. The long jumper's center of mass is within his hip region throughout most of the jump, but shifts forward, into open space, just before landing.

In high jumping, the ideal lay-out position, no matter what the jumping style, is one in which the largest possible amount of the jumper's body mass is below the crossbar at the peak of the jump. (See Figure 8.) In other words, some jumping styles allow the jumper's center of gravity to pass closer to the bar than others. (A jumper using a western roll must project his center of mass some 6 feet, 6 inches in the

Figure 8. In both the straddle and flop styles of high jumping, "draping" over the crossbar allows the jumper's center of mass to pass very near the bar.

air in order to clear 6 feet, while a straddle jumper might have to project his center of mass only up to 6 feet, 1 inch in order to clear the same 6-foot height.)

Theoretically, it is even possible to have enough body mass "draped" over the bar to allow the center of mass to pass under the bar while the jumper is going over it, but there is no evidence to date that this has actually occurred during any world-class jumps.

CHAPTER 4

Curves Of Flight

As mentioned in Chapter 2, an athlete or a throwing implement leaves the ground at an angle that is determined by two component velocities—one horizontal and one vertical. It is important, too, to know what happens to them after they leave the ground.

PARABOLIC CURVE

The moment an athlete leaves the ground and is free in the air, the entire "flight path" of his center of mass in the air is determined. This is especially noticeable in jumping and hurdling. The combination of forward-upward velocity when leaving the ground and the force of gravity causes the athlete's center of mass to follow a perfectly regular curve called a *parabola,* or *parabolic curve.*

When the athlete leaves the ground, the angle of his trajectory (and his curve of flight) has been determined by the combination of his horizontal and vertical velocities. (See Chapter 2.) During the jump, the horizontal component is unaffected by outside forces (except for some wind resistance), but gravity gradually slows the vertical component to zero and then reverses the process, causing the body to fall at exactly the same angle and velocity as it left the ground. The result is a perfect parabolic flight curve.

The depth (distance from take-off to landing) of the parabolic curve is determined by approach speed; the height is determined by take-off spring. Thus, the high jumper achieves a high, shallow parabola; the hurdler and long jumper achieve low, long parabolae. (See Figure 9.)

Figure 9. The height and depth of the parabolic curve is determined at take-off by approach speed and take-off spring.

It is a law of physics that no movement of the athlete, once he is free in the air, can alter the flight path of his center of mass. Changes in body position can change the position of the center of mass within the body during the time in the air, but wherever the center of mass happens to be in the body, and no matter how much it changes position, the center of mass must follow a perfect parabola.

In high jumping, it is obvious that the jumper's center of mass must be projected high enough at take-off to get the center of mass over the bar, and that the peak of the parabola be over the bar. No matter how efficient the jumper's bar clearance style might be, it is worthless if the take-off has not been good.

Some high jump lay-out styles allow the jumper's center of mass to pass closer to the bar than others. The important point in bar clearance is that as much body mass as possible be below the bar at the peak of the jump, so that

Figure 10. The long jumper's center of mass, which follows a predetermined parabolic curve in flight, shifts backward in the body when the arms are thrown back just before landing. Since the center of mass continues along that parabolic flight path, the legs have to shift forward.

the center of mass need not be projected quite so high at take-off in order to clear the height.

In long jumping, throwing the arms backward just before landing in the pit can add as much as five inches to the jump. (See Figure 10.) With the arms held back, the jumper's center of mass moves backward in his body, and thus the entire body position moves forward, since the center of mass is following its parabolic curve.

In hurdling, there is a slight dynamical advantage in lowering the head after leaving the ground. (See Figure 11.) Lowering the head lowers the center of mass in the body, which means that the entire body is raised slightly as the center of mass follows the pre-determined parabolic curve.

Figure 11. The hurdler's center of mass, which follows a pre-determined parabolic curve in flight, shifts downward in the body when the hurdler, in flight, lowers his head. Since the center of mass continues along that parabolic flight path, the hurdler did not have to jump quite so high (or spend quite as much time in the air).

The hurdler who lowers his head during hurdle clearance does not have to project his center of mass quite as high and does not have to spend quite as much time in the air.

The throwing implements must also obey the laws of physics. After being released, shots and hammers follow perfect parabolic curves, but disci and javelins, which are affected by aerodynamic laws, do not.

OPTIMUM ANGLES

The optimum angle for the projection of a missile is 45 degrees—if the point of landing is at the exact same level in altitude as the height of the release. (To achieve the 45-degree angle, the force of the vertical and horizontal components must be exactly equal. See Chapter 2.)

However, because all of the throwing implements are released above ground level, the angle of release must be less than 45 degrees. How much less depends upon the height of the release, the velocity of release, and, in discus and javelin throwing, on the aerodynamic properties of the implements.

In shot putting, the optimum release angle is between 40 and 42 degrees. The angle can be plotted by bisecting the angle formed by a line drawn from the shot at release to the eventual landing point and a vertical line drawn through the shot. (See Figure 12.)

In hammer throwing, because the hammer is swung close to ground level before being released, the angle of projection is only slightly below 45 degrees.

Because of aerodynamic forces, the discus and javelin do not follow perfect parabolic curves in flight. For this reason, the release angles are even lower than in the "parabolic throws" (shot put and hammer throw). The optimum release angle for the discus is between 35 and 40 degrees and for the javelin is between 30 and 35 degrees.

Figure 12. The ultimate release angle in shot putting depends upon the height of release and how far the shot will travel before landing. To plot the angle exactly, draw a vertical line through the release point and another line from the release point to the eventual landing point. By bisecting the angle formed by the two lines, the exact ultimate angle can be determined.

In long jumping, the athlete begins with great run-up and take-off speed, but he does so at the expense of vertical speed. The approach is so fast that there is nothing that can be done at the take-off board to add vertical velocity that compares with the horizontal velocity, so the angle of take-off must be considerably below 45 degrees. In actuality, it is seldom above 25 degrees.

To achieve the long jumping angle, the horizontal component is determined by the approach velocity of the jumper, of course. The vertical component is determined by the slight lowering of the center of mass during the final three strides, then raising it during take-off.

The movements which contribute to the vertical component in long jumping must not be so great, however, that they retard forward speed at take-off. If a jumper decides that he would like to get more height (as a coach might encourage him to do) by "gathering" more, he must slow down at take-off (perhaps unnoticeably) in order to do so. Slowing down decreases the length of the jump considerably. As an extreme example, a long jumper with 9.3 speed who decides he will raise his center of mass four feet at take-off (as in good high jumping), must slow down so much that the maximum that he will be able to long jump is 16 feet! Height is not the important factor in long jumping; continuing forward speed, right off the end of the board, is.

AERODYNAMIC FORCES

Implements affected by air resistance, such as the discus and javelin, do not follow a parabolic curve when thrown. Therefore, discus and javelin throwers must not only consider release angle and release velocity, but must also take into consideration the attitude angle of the implement, as well as wind direction and wind velocity.

CHAPTER 5

Axes

An axis is an invisible straight line passing through a turning body. Anything that turns, whether it is on the ground or in the air, turns around at least one axis, and thus has rotary motion. All parts of the body outside that axis turn at right angles to the axis.

When the athlete is turning while in contact with the ground, and the rotation is in a vertical plane, such as in the release of a javelin when the foot is planted and the upper body continues forward, the axis passes horizontally through the point where the foot meets the ground. (See Figure 25, page 65.)

When the athlete is in contact with the ground and the rotation is in a horizontal plane, such as the discus throw turn, the axis passes generally vertically through the point of support and the athlete's center of mass. (See Figure 13.)

When the athlete leaves the ground and is free in the air, all primary axes (those around which the entire body mass rotates) must pass through the body's center of mass. The primary axes, which are mutually perpendicular, are the longitudinal (or long) axis, which runs from head to toe; the frontal axis, which runs from side to side; and the saggital axis, which runs from front to back. (See Figure 14.) (The saggital axis seldom comes into play in the track and field events.)

Any object spinning in the air, be it inanimate object or athlete, must turn around its center of mass. For example, try spinning a carpenter's hammer in the air. Because the

Figure 13. When the athlete is in contact with the ground and rotating in a horizontal plane, the axis passes through the athlete's point of support and center of mass.

Figure 14. When the athlete is free in the air, the primary axes —longitudinal, frontal and saggital—are mutually perpendicular and all pass through the center of mass.

hammer's center of mass is in the head, far from the center of the hammer, the handle will rotate around the head as it turns in the air.

Knife throwers have knives with the weight in the handles so that they will always stick. As a knife hurtles toward its target, spinning rapidly around its center of mass, the blade is always closer to the target than the handle is. When the handle is in position to hit first, it is not close enough to the target. The blade, coming around, then sticks before the handle can get any closer.

Any of the primary axes in the body may be an *axis of momentum,* an *axis of movement,* or both, depending if the athlete is on the ground or not when the rotary action is initiated.

AXIS OF MOMENTUM

When rotation around an axis is begun while the athlete is still on the ground, and all the body mass outside the axis continues rotating around the axis after he is in the air, it is an axis of momentum. For example, the straddle high jumper's body rotates over the crossbar. (See Figure 15.) The jumper's longitudinal (head to toe) axis is the axis of momentum in this case.

AXIS OF MOVEMENT

When, after the athlete is free in the air, a movement is initiated around an axis (which must cause an equal and opposite reaction around the same axis) it is an axis of movement. For example, the pole vaulter is draped over the crossbar and suddenly raises his arms to clear the bar. His legs

Figure 15. The high jumper's roll is around an axis of momentum.

react by going backwards and up. (See Figure 16.) The vaulter's frontal (side to side) axis is the axis of movement in this case.

One excellent example of both an axis of momentum and an axis of movement is the frontal axis in the flop style of high jump bar clearance. The flop is, in effect, a back dive over the bar. The frontal axis, through the hips, serves as the axis of momentum as the athlete rotates, head first, over the bar. If the flopper were to continue in that position until landing in the pit, without bending at the hips, he would land

Figure 16. The pole vaulter's fly-away is around an axis of movement.

in a dangerous head-down position. However, the flopper bends at the hips and the frontal axis suddenly becomes both an axis of momentum and an axis of movement. The upper body remains in the same relative position on the way into the pit, allowing him to land on his back. And the equal and opposite reaction around the frontal axis is the raising of the legs to clear the crossbar. (See Figure 17.)

Figure 17. The flopper's frontal axis is both an axis of momentum and an axis of movement.

Nutation is the motion of the human body (or any object) in the air, when the rotation is around *two* primary axes at the same time. It is the result of an eccentric (off-center) thrust at take-off (or at release) which causes the rotation to form a conical pattern in the air. (See Eccentric Thrust, page 61.)

Nutation is purposely applied in acquiring proper lay-out and roll in straddle high jumping. It is often accidentally applied (and definitely unwanted) in an eccentrically thrown discus, javelin, or football.

CHAPTER 6

Momentum

Everything that moves has a certain amount of momentum, depending upon its mass and its velocity. (Mass times Velocity equals Momentum.) A 200-pound sprinter running 10 feet per second, for example, would have 2000 units of momentum.

If a wagon of bricks with a total weight of 100 pounds were rolling downhill at a steady speed of 30 feet per second (disregarding friction for the moment), that wagon would have a continuing momentum equaling 3000 units (mass of 100 pounds x velocity of 30 f.p.s.). If, suddenly, as the wagon continues downhill, 40 pounds of bricks were lifted from the wagon, reducing the total weight to 60 pounds, the wagon's momentum would not change, but its speed would immediately increase to 50 feet per second. By decreasing the mass, the velocity must increase so that the product of the two remains at its original 3000 units of momentum.

If two sprinters, one weighing 100 pounds and one weighing 200 pounds, are in the starting blocks side by side when the gun is fired, the 200-pounder must exert twice the force exerted by the lighter sprinter, just to stay even. If they both exert the same amount of force, the 100-pounder will accelerate twice as fast as the heavier sprinter.

Interestingly, if the 200-pound sprinter were to reduce his weight to 180 (without reducing his strength), he would be 11 per cent faster out of the blocks than when he weighed 200 pounds.

Conversely, weight men must rely on great mass for success in their events. Since mass times velocity equals momemtum, and only a certain amount of velocity increase is possible through training, the greater the weight man's mass, the better chance he has for success.

The deduction is, of course, that sprinters should lose weight and weight men should gain weight.

Football coaches have been known to say that a "small, tough football player with desire" can move a big football player, if he wants to badly enough. Since total momentum is what determines which player is "moved," the "small, tough" player must be able to generate more momentum than his opponent in order to move him. And since the opponent has more mass, the small player must make up for lack of size with superior speed. The problem is that in most cases the speed has to be more superior than is humanly possible.

A 150-pounder traveling at 30 feet per second (world class sprinting speed), for example, would be knocked flat if met by a 200-pounder traveling only 24 feet per second. "The bigger they are, the harder they fall" may be praiseworthy as a psychological gimmick to inspire small football players, but in practice it simply isn't true.

The following chart shows units of momentum generated for various weights at various speeds:

UNITS OF MOMENTUM

MASS IN POUNDS

VELOCITY IN FEET PER SECOND AND MPH	50	75	100	125	150	175	200
6 / 4.1	300	450	600	750	900	1050	1200
12 / 8.2	600	900	1200	1500	1800	2100	2400
18 / 12.3	900	1350	1800	2250	2700	3150	3600
24 / 16.4	1200	1800	2400	3000	3600	4200	4800
30 / 20.5	1500	2250	3000	3750	4500	5250	6000
36 / 24.5	1800	2700	3600	4500	5400	6300	7200

To demonstrate how important weight is to the football player, a weight gain of 20 pounds (provided the player does not slow down in the process) adds more to his momentum than a speed increase of from 11 seconds flat to 10.5 for 100 yards.

CHAPTER 7

Inertia

If a semi truck and a sports car decided to race away from a stop light, we know the semi would lose the race. The two vehicles could exchange engines, but it wouldn't matter. The semi would still lose.

Once the two vehicles were traveling down the highway at top speed, we also know it would take the truck a much greater distance to stop. The truck must start slowly and stop slowly because of its great inertia.

Inertia is a body's resistance to change in motion. As mentioned earlier, it is mass which causes the resistance to an increase in linear motion (as was the case with the heavy sprinter), and it is that same mass that resists the slowing down at the end of a race. This resistance is *inertia.*

When the motion is rotary (around an axis), the resistance to change in motion is dependent not only upon the amount of mass, but also on its distribution around the axis. The closer the mass is to the axis, the less the resistance to speeding up or slowing down; the farther the mass from the axis, the greater the resistance.

This resistance to change in rotary motion is called its *rotary inertia,* and is found by multiplying the rotating mass times the radius of rotation, squared (mr^2=r.i.). The mass

(m) in the formula is the entire weight that lies outside the axis of rotation; the radius (r) in the formula is the distance from the axis to the center of that rotating weight.

The inertia of a 16-pound shot is twice that of an 8-pounder. It is because of the 16-pound shot's greater inertia that it cannot be put as far as the 8-pound shot, with the same amount of force.

Rotary inertia is very important in discus throwing. The exact size, shape and weight of a discus are determined by the rule books, but the distribution of that weight is *not,* and that is how some discus throwers have a decided advantage over others. They are using disci that have the weight distributed around the outer edge, with very little weight in the middle.

When a discus is thrown, the thrower applies both translational and rotational kinetic energy to the discus—translational energy for the distance and rotational energy for the spin.

A hollow discus with the weight distributed to its outside has a greater rotary inertia than a "solid" discus. Extra effort must be imparted at release to get it spinning, but once the discus is turning in the air, it maintains its gyroscopic effect and stays level in the air much longer—adding as much as 7½ feet to a 150-foot throw.

CHAPTER 8

Conservation Of Rotary Momentum

Rotary momentum is rotary velocity (the speed of turning) times the turning body's rotary inertia (resistance to turning).

To better visualize rotary momentum, consider a pirouetting ice skater, spinning at a very high rate of speed. The skater's rotary momentum is the velocity of turning times a turning "center of mass," which is a complete circle of points, about waist high, somewhere between the turning axis and the outermost extremities of the skater. The size of this circle of points is determined by multiplying mass times the radius, squared (mr^2). The exact position of this circle is not important; the important thing is that rotary momentum is constant, and by extending his arms away from his body, increasing the size of the circle (increasing rotary inertia), the skater may reduce rotary velocity, and vice versa. This concept, called *conservation or rotary momentum,* is one of the most important in the analysis of sports movements.

A turning body, free in the air, has a constant rotary momentum. This rotary momentum must begin on the ground and cannot be added to or subtracted from once the body is in the air. It remains the same until the body returns to the ground.

A straddle high jumper, for example, does not increase his rotary momentum over the bar by suddenly twisting his body as he goes over (although many jumpers think they do). Rotary velocity over the bar (the speed of the jumper's roll) may be increased only by decreasing rotary inertia (by shortening the turning radius). Since the product of rotary velocity and rotary inertia is rotary momentum (a constant, once free in the air), reducing one factor must increase the other. In other words, when the high jumper pulls his arms and legs in close to the axis of rotation as he clears the bar, his rotary velocity must increase greatly momentarily, only to be reduced again as he stretches out for the fall into the pit. (See Figure 18.)

Simply stated, then, shortening any turning radius increases speed. Divers and trampolinists have known this for years, but only recently have track and field athletes begun to realize the importance of this concept.

The sprinter reduces the rotary inertia of his recovery leg as it comes forward, by bending the leg as much as possible at the knee. The leg moves forward faster, contributing to an increase in sprint speed. (See Figure 19.)

Also, a runner's arms may be swung forward and backward more rapidly (and with less energy) when they are bent at the elbows.

A pole vaulter who wants to delay his pull on the pole keeps his body reasonably straight as it swings forward. But when he shortens his body by bending at the hips and bending his knees, he is able to quickly swing his body forward and up into the push-off position. (See Figure 20.)

A figure skater is able to spin so rapidly on the ice

Figure 18. The high jumper's speed of turning over the crossbar increases because of the great decrease in rotary inertia—the resistance to turning.

Figure 19. The runner's bent recovery knee increases rotary velocity because the bending of the knee decreases rotary inertia.

Figure 20. The pole vaulter shortens his body and swings up much more easily.

that his movements become a blur to the human eye. He does this by keeping his arms and legs wide as he begins the spin, and then pulls arms and legs in close to the long axis, greatly increasing rotary velocity. (See Figure 21a.)

Then, when the skater wants to come out of the spin, he stretches his arms and legs out wide, increasing the rotary inertia, stopping his spin and taking a bow, all at the same time. (See Figure 21b.)

The same principle may be used in discus throwing, although few American discus throwers are taking advantage of it. By starting wide, with arms and legs as far from the long axis as possible, and then shortening rotary inertia in the center of the circle, the speed of the turn (and, thus, the speed of the release) may be increased greatly. (See Figure 22.) Since of the three factors contributing to length of throw (speed of release, angle of release, angle or tilt of discus) the only factor which may be continually improved upon is speed of release, this should be the primary consideration of the coach.

The discus thrower's upper body, during the turn, has a greater rotary inertia than the lower body, because of the mass of the discus at the end of the outstretched arm. This slows the upper body and keeps it trailing the lower body throughout the turn, with the body ready to be unwound for greatest effect at the time of release. The discus must trail along behind during the turn. A coach who instructs a discus thrower to "keep the discus back" during the turn is wasting his time. The discus cannot possibly be kept any place except behind, because of the great rotary inertia.

In sports, there are many examples of the use of conservation of rotary momentum to speed up or slow down

Figure 21. The pirouetting figure skater is able to spin rapidly because of the law of conservation of rotary momentum.

Figure 22. The discus thrower's turning speed at the time of release can be increased greatly by starting wide, creating great rotary inertia, and then reducing that inertia before the release.

turning motion.

The twisting trampolinist stretches his arms out to the sides to stop twisting before returning to the trampoline bed.

Pole vaulters and flop-style high jumpers, if worried about landing on their heads, may automatically raise their arms above their heads when falling, to slow down the turning movement around the frontal axis.

The hang-style long jumper hangs in the air, without tucking until the last moment, to retard forward rotation around the frontal axis.

Conservation of angular momentum may be easily

demonstrated on a freely-turning turntable, such as a piano stool. Have someone stand on the turntable, holding a heavy object (such as a 12-pound shot) in both hands, in front, at arms' length. Then, turn him very slowly. As soon as the demonstrator is turning on his own, have him pull the shot in to his chest. To the surprise of the demonstrator, and those watching, the velocity of turning increases greatly. As soon as the shot is thrust back out to the original position, the turntable slows immediately to the original turning speed.

The more mass (the heavier the object) held away from the body when the turning motion begins, the greater the velocity when rotary inertia is reduced by pulling the mass in close to the axis.

Another way to demonstrate is with a ball on a six-foot string. Hold the end of the string and swing the ball around the head until a certain momentum is established. Then let the string shorten by wrapping around your arm until the ball finally hits the arm. Using mr^2, when the string is six feet long, rotary inertia is 36m and the velocity of turning is x. When the string length reduces to three feet, or 9m, the velocity increases to 4x. When the string length shortens to one foot, or 1m, the rotary velocity increases to 36x, or 36 times the original speed. Throughout, rotary momentum has remained unchanged. Rotary velocity increased because rotary inertia (the resistance to change) decreased.

In a hypothetical case, if the turning momentum of a human body in the air were 200 units, the turning velocity were 2 revolutions per second, and the radius of turning were a distance of 10, we would have:

Rotary Momentum = Rotary Velocity × Rotary Inertia (mr^2)
200m   =   2   × $m \cdot 10^2$
200m   =   2   ×   100m

If, then, the radius were cut in half, by pulling arms and legs in close to the body, we would have:
200m   =   8   × $m \cdot 5^2$
200m   =   8   ×   25m

By reducing the radius to half the original, rotary velocity has been increased *by four times!*

From this example you can see that a slight reduction in rotary inertia can bring about a proportionately greater increase in rotary velocity. This explains the twisting movements that divers and trampolinists are able to accomplish, when seemingly getting no rotation at take-off. A very slight rotation may be greatly accentuated in the air by pulling all the extremities in close to the turning axis when desiring a fast twisting movement. (See Figure 23.)

Figure 23. Trampolinists have learned (usually through trial and error) that great twisting can be accomplished by pulling in the limbs, greatly decreasing rotary inertia.

CHAPTER 9

Turns From The Ground

In almost all sports movements, athletes leave the ground, at least momentarily, and turn briefly around an axis of momentum. The rotary momentum the athlete has in the air must come from the ground, and must be begun in one or a combination of the following three ways:

    1. Through eccentric thrust
    2. By checking linear motion
    3. Through transference of rotary momentum

Eccentric thrust comes about at take-off when the resultant line of force from the ground does NOT pass directly through the body's center of mass.

If the line of force were to pass *directly* through the center of mass (as when a basketball center goes up for the opening tip-off), there would be no rotation. The athlete goes straight up, comes down in the same place, and his body is in the same relative attitude on landing as it was at take-off.

Eccentric thrust creates rotary momentum, but it also must reduce effective force. In fact, the more the eccentric thrust at take-off (and thus the more the rotary momentum around one or more axes), the less the effective force (and thus the less height or distance attained).

Eccentric thrust is the primary means for achieving

the lay-out position and the rotation over the crossbar in straddle roll high jumping. The beginning straddle jumper learns to lay out and to roll over the bar in his first practice sessions. He does not know it, but he is teaching himself eccentric thrust at take-off. If he were to drive upward on an exact line from his take-off foot on up *through* his center of mass (thus with no eccentric thrust), his center of mass would attain maximum height, but his body would not rotate into a lay-out position.

In order to get the rotation necessary for lay-out, the straddle jumper must lean toward the bar at take-off, driving upward from the take-off foot along a line that is not through his center of mass. (See Figure 24.) The farther away from the center of mass this imaginary line is, the more eccentric thrust and rotation in the jump, and the less height achieved in the jump.

Therefore, straddle jumping becomes a compromise. Getting enough rotation for bar clearance requires eccentric thrust, which reduces the height of the jump. (It should be emphasized again that the technique of thrusting eccentrically is developed by the straddle jumper automatically in practice sessions. He does not realize that he is doing it; nor does he know it is curtailing his "height-getting ability.")

By the same token, in baseball pitching, a curve ball can never be thrown as fast as a fast ball, no matter how much the pitcher might want to throw a fast curve. The eccentric thrust required at delivery to spin the ball and make it curve must reduce the effective force of the pitch (and the speed of the ball). The more the spin, the slower the pitch.

Similarly, the more rotation a diver gets from the div-

Figure 24. The jumper on the left is jumping more eccentrically than the jumper on the right, and will not jump as high as he would otherwise. He does this, obviously, because he has not perfected the action-reaction method of bar clearance (See Chapter 10); he must rely on rotation from the ground to roll him over in the air.

ing board, the less height achievable. The diver cannot go nearly as high executing a triple somersault as he can doing a jackknife.

Using an ordinary pop bottle, there are two good ways to demonstrate eccentric thrust:

First, balance the bottle on its side on one finger and toss the bottle straight up, a foot or two. (Be sure to catch the bottle!) Then, holding the bottle up in the same position with the opposite hand, use the same finger to toss the bottle up with the same force, but this time apply the force near the end of the bottle instead of under its center of mass. Rotation is evident and height is decreased.

Second, with the bottle on its side, toss it into the air, rolling it off the hand so that the bottle rotates around its long axis. When the upward force is directly through the bottle's center of mass, there is no wobble. When there is eccentric thrust, the bottle rotates around *two* axes and prescribes a conical path in the air. (See Nutation, page 45.)

For this same reason, a discus must be thrown with some eccentric thrust to attain the necessary stabilizing spin in the air, but there will always be a wobble in the air whenever the thrower applies the force either above or below the plane of the discus' center of mass, rather than beside it.

A football, too, must be thrown with eccentric thrust to attain necessary stabilizing spin, but the way a football must be thrown makes it impossible to completely eliminate the wobbling motion. Even on the best throws there will always be a wobble.

Again, most athletes do not relate force and center of mass with wobble. They know they don't want a wobble, and so, through trial and error, they develop the technique of applying force as close to the center of mass as possible, so there is very little wobble.

Checking linear motion. When the linear motion (motion along a straight line) of a rigid object is interrupted at one end of the object, the result is forward rotation. The other end of the object continues ahead, but at an increased rate of speed, and rotation has begun. A passenger jumping off a fast-moving bus, for example, demonstrates the principle of checking linear motion. As he lands he must either be able to run as fast as the bus is moving, or fall forward to the ground.

When a javelin thrower plants his front foot prior to releasing the javelin, his upper body and arm (and, of course, the javelin) increase in forward speed, aiding in increasing the javelin's velocity at the time of release. (See Figure 25.) The

Figure 25. The javelin thrower checks linear motion when he plants his foot, and the upper body, arm and javelin continue on at an increased rate of speed.

axis of rotation passes through the point where the foot meets the ground.

The flop-style high jumper checks linear motion to achieve his lay-out position over the crossbar. Linear motion is checked when the take-off foot is planted, and the upper body continues to travel forward, giving the body the rotation necessary for achieving the lay-out position. (See Figure 26.)

Unlike the eccentric thrust necessary to bring about needed rotation in straddle jumping, checking linear motion in the flop style does not check "height-getting" ability ap-

Figure 26. The flopper checks linear motion by planting his take-off foot. The upper body continues on at an increased rate of speed.

preciably. Therefore, it can be said that the flop is a "less inefficient" style of high jumping than the straddle roll, since neither style is totally efficient.

It may be that someday the principle of stopping linear motion will be employed to increase the world record in the long jump. The long jump, like straddle high jumping, is a compromise—the athlete must give up a portion of one advantage in order to gain some other advantage that is more beneficial.

As was mentioned earlier, in the straddle high jump the jumper gives up some height, through eccentric thrust, in order to lay out and roll over the bar. In the long jump the jumper must actually decrease forward speed slightly at take-off, in an unconscious effort to reduce forward rotation and assure a more economical landing position.

It would seem possible, through the study of dynamics, that the long jumper would not have to reduce forward speed at take-off if he would do a complete somersault from the take-off board into the pit, taking maximum advantage of forward rotation. (Don't laugh. Remember that Fosbury's high jumping style was considered freakish until the 1968 Olympic Games!)

There are three ways the long jumper of the future may use the somersault technique to set new records:

    1. He may begin to turn around his longitudinal axis at take-off, Fosbury style, and do a back somersault into the pit.

    2. He may do a straight forward somersault, landing in the pit in approximately the same position as present-day jumpers.

3. He may do a forward somersault with a half twist, so he can land backwards in the pit.

There are dynamical disadvantages in each of these styles, but all seem to be better, on paper, than any style being used at this writing. The somersault style—someday in the future—should be good for jumps of 30 feet or more.

When the somersault style is finally used (probably by a gymnast turned track man), the question will arise as to the legality of this new jumping style. And the officials will have to decide, subjectively of course, whether the somersault style should be allowed.

Some gymnastic improvements in track and field events have been allowed in the past and some have been banned. Fiberglass pole vaulting and the Fosbury flop are allowed; Dickie Browning's world record-breaking back somersault high jumping style (technically from both feet) was disallowed, as was the discus-like spinning javelin style developed by two clever Spaniards, who broke the world javelin record a few times before the rules had to be changed. But one thing is certain: No one will venture an opinion on the somersault long jumping style until an athlete does well with it!

Transference is another means of creating rotary momentum in the body at take-off, and for adding impetus to jumps and throws. Momentum may be transferred from one object to another (as when a bowling ball hits the pins), from the entire human body to one part of the body, or, most commonly, from one part of the human body to the entire body.

An athlete lying on his back who wants to sit up may

lift his legs until his feet are directly above his head, and then snap his legs downward, suddenly locking them there. The momentum of his moving legs transfers to his entire body, bringing the athlete into a sitting position.

From there the athlete might want to get up to his feet. So he swings both arms upward and then locks them. The momentum of his swinging arms transfers to his entire body, aiding the legs in bringing the athlete to his feet.

Listed here are a few of the many examples of transference of momentum in track and field:

The straddle high jumper transfers momentum from the straight lead leg and the lifting arms to the entire body, aiding in vertical acceleration.

The long jumper, on landing, moves his arms suddenly forward, keeping the body from sitting back in the pit.

The flop-style high jumper transfers momentum from his leading knee to his entire body to achieve the back down position over the crossbar. (See Figure 27.)

The discus thrower transfers momentum from the entire turning body to the throwing arm and discus by stopping the turn and allowing the arm and discus to continue. (See Figure 28.)

Some basketball coaches instruct their players to keep their hands and arms above their heads when rebounding a free throw, so they won't waste any time bringing the hands up. In reality, however, those players would be able to jump as much as a foot higher, through the transference of momentum, if they would start with the arms low and then swing them up as they went for the rebound.

Figure 27. The flopper's knee swings up and across the body just before take-off and the momentum transfers to the entire body.

Figure 28. The discus thrower uncoils rapidly and then stops suddenly. The body's momentum is transferred to the upper body, arm and discus.

CHAPTER 10

Turns In The Air

When an athlete attempts to turn his body while he is free in the air, without first obtaining some rotation from the ground, the only movement he can engender is an initiated action and an equal and opposite reaction. Newton's 3rd Law of Motion is as infallible today as it has been since the beginning of time: *For every force there is an equal and opposite force of reaction.* It is impossible to create rotary momentum once the athlete is off the ground.

To demonstrate, jump straight up in the air (without rotating from the ground) and at the peak of the jump, swing the arms in one direction. The legs will immediately cross in the opposite direction. If this is not graphic enough, do it again, this time throwing a medicine ball. (See Figure 29.)

For even more complete experimentation, stand on a turntable (which simulates being free in the air) and turn the upper body in one direction. The lower body must turn in the opposite direction. (See Figure 30.) Action and reaction occur in the same plane, or in planes that are parallel, around an axis which passes through the body's center of mass.

Even though Newton's 3rd Law is scientifically indisputable, there are still coaches who "break the law" by teaching their straddle high jumpers to hook the leading arm down

Figure 29. Action-reaction can be demonstrated by jumping straight up (without getting any rotation from the ground) and throwing a ball to one side.

and under the crossbar during bar clearance, which can only aid in knocking the crossbar off.

As the straddle jumper assumes his lay-out position over the crossbar, the long axis through his body, which parallels the crossbar, is an axis of movement. For every rotational action around that axis on one side of the body's center of mass, there must be an equal rotational reaction in the opposite direction on the opposite side of the center of mass. (See Figure 31.)

Figure 30. Action-reaction is most graphically demonstrated on a turntable, which simulates being free in the air.

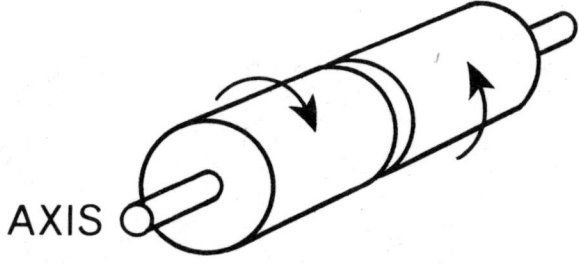

Figure 31. In straddle jumping, action and reaction occur in planes that are on opposite sides of the center of mass, but are parallel to each other.

In other words, if the jumper raises his trailing leg, he must also raise his leading arm. (See Figure 32.) Lowering his leading arm can only lower the trailing leg.

One reason it has been so difficult convincing straddle jumpers and their coaches that raising the lead arm is necessary for bar clearance is that all good jumpers do it, but it is difficult to see with the naked eye. The action is so fast the observer must concentrate on watching the lead arm, disregarding everything else, before the movement can be seen. Many straddle jumpers who raise the leading arm during bar clearance (including some world-ranked Americans) claim they don't—until they study photos of themselves in action.

It is usually difficult to determine which is the cause

Figure 32. The straddle jumper raises his lead arm in order to clear his trailing leg.

and which the effect during the action and reaction of straddle bar clearance. Those jumpers who are aware of the importance of raising the lead arm during clearance probably do so consciously. Those who are not aware concentrate on raising the trailing leg, and the reaction forces the lead arm up.

At the lower heights it is possible for the straddle jumper to clear the bar without raising the trailing leg, and, in such cases, the leading arm is not raised. The jumper does this by increasing his body rotation at take-off (which decreases his height-getting ability).

When the jumper approaches his maximum heights, however, he must raise the trailing leg during bar clearance, and the leading arm must also be raised. If the leading arm is not allowed to raise, or is forced down (as some coaches instruct their jumpers), the trailing leg cannot be raised to clear the bar.

There is an important action-reaction movement in the flop style of high jumping, too. As the legs are brought up to clear the crossbar, the equal and opposite reaction in the upper body is the maintaining of the relative position of the head and shoulders, which had been rotating toward the pit. (See Figure 33.)

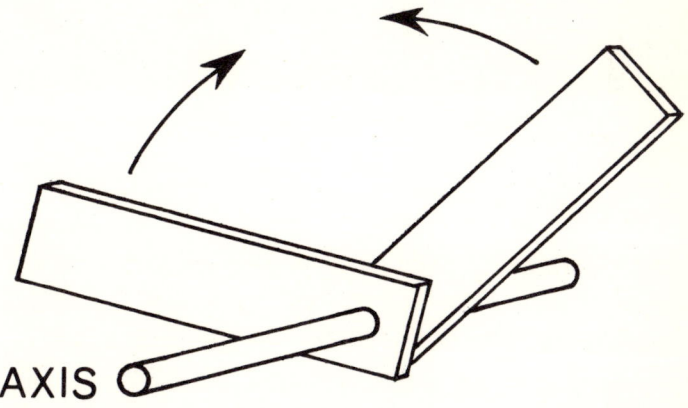

Figure 33. In the flop, action and reaction occur in the same plane, facing each other.

The frontal axis (through the hips) is still an axis of momentum, but it has also become an axis of movement, with action on one side of it and reaction on the other. This action (and the simultaneous reaction) are especially beneficial in this case, since they serve a necessary dual purpose. Raising the legs gets the legs over the crossbar; the reaction to this keeps the jumper from landing on his head.

In the hurdle races, action and reaction play a very important role. When a coach tells his hurdler to lean his trunk farther forward as he clears the hurdle, he can only do this if he also raises his lead leg higher, for the two cannot be separated. One (which one doesn't really matter) is the action and the other is the equal and opposite reaction. (See Figure 34.)

As the high hurdler's trailing leg comes forward over the hurdle, the equal and opposite reaction is the backward movement of the lead arm. (See Figure 35.) If the hurdler is to remain facing straight ahead throughout his flight over the barrier, the action and the reaction must be exactly equal. But the leg has more mass than the arm, and so the arm must swing out wider than the leg (sometimes even into the next lane) in order for the two to be perfectly balanced. Too often a coach will insist that his hurdler keep that lead arm in close as it goes back, giving the hurdler two choices as he learns hurdling skill through trial and error: 1. He can land off balance, making it difficult (or impossible) to make the next hurdle, or 2. He can learn to take off off-balance in order to land correctly. Obviously, neither is more desirable than allowing the arm to swing wide. (The double arm lead, which some coaches still teach, can only make things worse.)

At least one world class high hurdler solved the problem by reaching his lead arm in front of his body as he cleared each hurdle, instead of reaching his lead arm straight ahead, so he could balance the action of the trailing leg by swinging the lead arm back through a wider arc. (See Figure 36.) This seems to be the perfect answer to the balance problem in hurdling.

Figure 34. Once the hurdler is free in the air, the lead leg and trunk are "tied together" by action-reaction. One cannot move without the other.

Figure 35. As the hurdler's trail leg moves forward, the lead arm must move back, equally. If the arm is held in so that its rotary inertia is not equal to that of the trail leg, then the equal and opposite reaction will have to include more of the body and the hurdler will land off-balance.

Figure 36. By getting the arm in a position to react through a longer arc, the hurdler can solve his action-reaction problems.

CHAPTER 11

Inertial Axes

When Newton's 3rd Law (action-reaction) was mentioned earlier, it was shown that a rotary action around the long axis in one part of the body (as in straddle high jump bar clearance) requires an equal and opposite rotary reaction in the opposite end of the body.

The number of degrees of the actual reaction depends upon the rotary inertia (resistance to turning) of the two halves of the body. For example, if the straddle jumper over the bar rotates his upper body through 30 degrees, and the lower body has twice the rotary inertia of the upper body (because of the greater mass of the legs and their more spread-out position over the bar), the lower body will rotate only 15 degrees in the opposite direction.

It is because of this principle that a cat (or rabbit, or astronaut, or trampolinist) is able to turn over in the air without having received rotation from the ground. (This concept never really comes into play in track and field, but it is mentioned here to answer any skeptics who might use inertial axis examples to discredit the "Rotation Must Come From the Ground" concept, which *is* true in the track and field events.)

To demonstrate how an athlete may take advantage

of an inertial axis in order to turn over in the air, it is best to use a rabbit. Cats have such fast reactions that it is impossible for the man holding the cat to release his hands before the cat has begun to rotate from his hands. Rabbits have much slower reactions, and are not able to begin to turn until they are free in the air.

When the rabbit is held upside down and is suddenly released, he immediately bends his body so that the long axis becomes two nearly perpendicular inertial axes, both passing through his center of mass. (See Figure 37.) (The secret of turning over is the ability to bend in the middle. Rabbits, cats and humans can do it; dogs cannot, and when dropped from an upside-down position are destined always to land on their backs.)

Instantly, the rabbit rotates his upper body (Axis A) through 90 degrees. Because of Newton's 3rd Law (See Page 72.), the lower body turns in the opposite direction, but the great mass outside Axis A in the lower body (caused by the bending) allows only a few degrees of reaction. Notice in Figure 37 that the reaction in the lower body has caused the rabbit's back paws to cross.

Almost simultaneously, rotation around Axis B follows rotation around Axis A. Because of great flexibility, the rabbit is able to twist around each inertial axis, using the great rotary inertia of the other end of his body to allow him to turn over.

Throughout the rabbit's fall, his rotary momentum remains at zero. Once the turning movement is over, the rabbit cannot continue turning. Nor can a diver or trampolinist use this technique to initiate a continuing twisting movement in the air. The action and reaction during the turning over

Figure 37. The rabbit (unknowingly) takes advantage of two inertial axes to turn over in the air.

process are simultaneous. When one stops, the other must also stop, and thus there can be no continuation of a twisting movement. This technique can be used, however, to bring about a change of body position while in the air.

If a diver were released from a very high diving board in a prone, back-down position, he would be able to bend at the waist during his fall, creating two inertial axes, and turn over before hitting the water. If, however, his waist were splinted so that he could not bend in the middle, he would fall, helplessly, all the way to the water in that back-down position.

Such is not the case in sky diving. Because speeds in excess of 100 m.p.h. are reached by sky divers, they can take advantage of the 100+ m.p.h. wind resistance, aerodynamically, and turn over in the air by moving an arm or a leg very slightly. But in track and field, where the greatest falling speed reached in any event is about 21 m.p.h.*, wind resistance cannot be considered a factor of importance.

Because of the inertial axis principle, it is easier to maintain so-called "good form" in high hurdling than it is in intermediate or low hurdling. As the trailing knee moves forward over the high hurdle, and the upper body reacts in an opposite direction, a great deal of rotary movement is neutralized by the inertial (vertical) axis.

The trailing knee moves forward, and its rotary action (around the vertical axis) moves effectively through approximately 180 degrees. The extreme forward lean of the hurdler's body and the position of the lead leg make the vertical axis an inertial axis, with great rotary inertia. (See Figure 38.)

---

*If a pole vaulter clears 18 feet and falls 15 feet onto a 3-foot pit, he is traveling at 21 m.p.h. when he lands.

Figure 38. The hurdler's inertial axis helps to absorb some of the action and reaction over the hurdles.

Because of the mass that is outside that vertical axis, some of the rotary action (and its reaction) is "absorbed" in the body.

In intermediate or low hurdling, where the lead leg is not lifted so high (and thus, because of action-reaction, there is far less forward body lean), very little of the rotary action of the trailing leg coming forward and its upper body reaction can be neutralized by this diminished inertial axis. In order for the hurdler to maintain balance, upper body rotary inertia must be increased by increasing the radius, by extending the lead arm and swinging it backward in a larger arc. The result is an apparent "flailing" of the arms in order to maintain balance. (See Figure 39.)

Figure 39. In low and intermediate hurdling, the arms must be swung wider in order for the hurdler to maintain balance.

CHAPTER 12

Falling Bodies

Contrary to popular belief, it does *not* require strength to keep the legs up prior to landing in the long jump. The coach can waste time having his long jumpers hang from chinning bars, holding their legs straight out, developing the strength necessary to hold their legs out when hanging from a bar, but that strength will have absolutely no effect on the so-called ability to keep the legs up just before landing in the pit. In fact, it takes as much effort for the long jumper to hold his legs *down* as it does to hold them up, prior to landing.

Until the time of Galileo it was believed that heavier objects fell faster than light objects. We now know, disregarding air resistance, that all objects, regardless of weight, fall at exactly the same speed.

If you were to drop a marble and a cannonball from the same height at the same time, the two would accelerate equally (at an increasing velocity of 32 feet per second every second), and they would hit the ground at the exact same time. If the marble and the cannonball were welded together and then dropped, side by side, there would be no change; they would still drop at the same rate of speed, with no change of relative positions.

An arrow shot into the air (or a bomb dropped from

an airplane) cannot be made to land point first by making the point heavier. It must have fins to catch the air, forcing the tail of the arrow (or bomb) to fall behind.

A javelin does not land point first because its point is heavier than its tail; it does so because the tail is designed to catch more air than the point, causing the tail to fall behind.

Similarly, if you were to hold a carpenter's hammer in a flat position and let it fall to the ground (from any height), even though the head of the hammer is considerably heavier than the handle, the hammer would not change position in the air.

The same is true of the human body free in the air— be it sky diver or long jumper. All parts of the body fall at the same rate of speed; the legs fall at the same rate as the rest of the body. (See Figure 40.)

Figure 40. All parts of the long jumper — body, arms, legs, etc. — fall at the very same rate of speed.

It is easy to see, then, that it could not possibly require any strength for the long jumper to hold his legs up, no matter what we've always heard. With no effort the legs can be held up as long as the jumper is free in the air. (That is not true, of course, when the jumper is not free in the air, such as when hanging from a chinning bar.)

The question is: What does cause the legs to appear to drop during the landing phase of the long jump? One answer is action-reaction; the athlete instinctively raises his upper body before landing, which also lowers his legs. (See Figure 41.) The other answer is *forward* rotation around the

Figure 41. When the long jumper raises his upper body before landing (perhaps hoping to stay in the air longer), his body opens like a jackknife around his frontal axis, and the legs drop prematurely. Good "leg extension" in the air is useless if it is not maintained until the landing.

frontal axis, which passes through the jumper's center of mass. The legs remain in the same relative position, but the entire body rotates forward around the center of mass (the result of checking linear motion at the take-off) so that the legs appear to be dropping prematurely. The real difference is the attitude of the entire body—not just the legs. (See Figure 42.)

Figure 42. Forward rotation, caused by checking linear motion at take-off, can put the body in a position so that it is impossible to raise the legs. In this case, not only have the legs dropped prematurely, but forward rotation has also caused the jumper to fall forward in the pit.

CHAPTER 13

Secondary Axes

When an athlete is free in the air and wishes to change his body position before landing, he may do so by creating movement around a secondary axis—an axis which does *not* pass through his center of mass.

For example, a ski jumper who is slightly off balance in the air swings his arms (and sometimes even his ski poles), windmill fashion, to regain his balance before landing. The secondary axis, created through the skier's shoulders, helps him change his body position in the air. (See Figure 43.)

Secondary axes may be employed while in contact with the ground, too. The beginning roller skater swings his arms, sometimes violently, to regain his balance.

The circus wire walker (if he doesn't carry a balance pole or parasol) must swing his arms to stay on the wire.

The sprinter who throws his weight forward and leans into the tape in a close finish may have to swing his arms, windmill fashion, in order to regain his balance and keep from falling, head first, on the track. (See Figure 44.)

The long jumper who hitch-kicks in the air may not realize how important that movement is to him—or that he is actually creating two secondary axes when he "runs in the air." The jumper approaches the board at near top speed,

Figure 43. The skier uses his arms to regain balance.

Figure 44. A pronounced lean at the tape can cause a runner to stumble and fall. The arms may be swung, however, to help regain balance.

with his upper body traveling at the same rate of speed as his lower body. But during the fraction of a second when the take-off foot stays on the board, the upper body continues forward (as much as four to five feet) and increases in speed. (See Figure 45a.) The result is unwanted forward rotation

Figure 45. The long jumper neutralizes his forward rotation by creating two secondary axes.

around the center of mass (caused by checking linear motion), which must be neutralized in some way during flight to keep the jumper from doing a nosedive into the pit.

Over the years, through trial and error, long jumpers have developed different ways of neutralizing the undesirable forward rotation that is inevitable following a good long jump take-off. By far the best of these techniques is the hitch-kick, which is really the creation of one secondary axis through the shoulders and one through the hips.

After leaving the board, the jumper's body begins to rotate forward around the frontal axis, which runs through the body from side to side, through the body's center of mass. (See Figure 45b.) To counteract this rotation, the jumper hitch-kicks. Each leg is straight as it moves backward and bent at the knee as it moves forward. The difference in the rotary inertia of the two legs during these movements causes the jumper's lower body to move forward. (See Figure 45c.)

For the greatest effect, the arms should be used in the same way. Each arm must be straight as it moves forward (windmill fashion) and bent at the elbow as it moves backward. The difference in the rotary inertia of the two arms causes the jumper's upper body to move backward.

By hitch-kicking and creating the two secondary axes, the jumper is able to delay forward rotation, enabling him to assume a better landing position. However, at the completion of the hitch-kick the original rotation continues again, and that forward rotation helps to create the illusion that the jumper cannot keep his legs up just before landing.

For many years it was believed that if the hitch-kick served any purpose at all, it was to maintain rhythm and balance in the air. Now we know it is the means for keeping the legs from dropping into the pit prematurely. (Again it should be emphasized that the hitch-kick, like most acceptable athletic techniques, was discovered by athletes, through trial and error, and only in recent years has its real value been realized.)

In triple jumping, a modified (legs only) hitch-kick is employed during the hop phase (and sometimes during the step phase) to delay forward rotation. During the jump phase, the jumper may also use a brief hitch-kicking motion, or, if he does not have enough time in the air, he may prefer to use the hang style.

CHAPTER 14

Action-Reaction On The Ground

All body movements from the ground are the result of ground reaction, a force equal and opposite to a force exerted by the athlete. In other words, the greater the force applied to the ground by the athlete, the greater the force given back to the athlete by the ground. This concept is a part of Newton's Third Law and is very important in track and field coaching.

If a man who weighs exactly 150 pounds stands on the ground, the ground reacts by pushing back with a force of exactly 150 pounds. If the man suddenly pushes against the ground with a force of 200 pounds, the ground pushes back with an equal force of 200 pounds and the man jumps into the air. The greater the force against the ground, the higher the jump.

In high jumping, ground reaction is by far the most important coaching consideration. The more force exerted against the ground, the higher the center of mass can be raised. It is possible for a 160-pound high jumper to increase his weight against the ground to over 600 pounds at take-off, but his jumping leg must have the strength to support 600 pounds for that fraction of a second.

A sprinter can increase his speed from the blocks by

increasing his force against the blocks. The two forces are equal and opposite. But, again leg strength is the determining factor.

An athlete's effective force (or weight) against the ground can be increased by swinging moving parts of the body away from the direction of force just before leaving the ground. In the sprint start, the leading knee and opposite arm help to increase force against the block. (See Figure 46.)

Figure 46. The sprinter's leading knee and opposite arm increase his force against the front block.

A high jumper lifts his arms and swings his free leg up straight, increasing his effective weight by almost four times. (See Figure 47.)

The high jumper's force against the ground can be increased even further by increasing the weight of the free leg as it swings upward. If the jumper wears a heavy shoe on his lead foot, or wears an ankle weight, (or, more subtly, wears a heavy lead insole), he will jump higher—*provided his jumping*

Figure 47. The high jumper's arms and lead leg increase his force against the ground.

leg is strong enough to support the additional weight. (Jumping with no shoe on the lead foot has an opposite effect, of course. But there are still jumpers who insist that they don't want to carry that extra weight over the bar.)

A shot putter exerts force against the shot, but he is also exerting force against the ground as he lifts the shot, and the ground reacts by giving an equal amount of force back to the putter. To exert *maximum* force in any throwing event, the athlete must be in contact with solid ground until the implement is free in the air. If the athlete is in the air when the implement leaves his hand, the body absorbs the reaction and no additional force can be imparted by the ground. (See Figure 48.)

In sprinting, two factors contribute to speed over the ground—stride frequency and stride length. (Stride Frequency x Stride Length = Speed.) Obviously, in order to improve speed, one of these two factors must be improved.

Strides Per Second		Length	Time for 100 Yards
4	x	6'8"	11.3
4	x	7'0"	10.8
4	x	7'4"	10.3
4½	x	6'8"	10.0
4½	x	7'0"	9.5

A stride length increase of four inches is equal to approximately five-tenths of a second in the 100-yard dash. Stride frequency increases of one-half stride per second are equal to about one and three-tenths seconds in the 100.

However, stride frequency is largely an inborn characteristic and, in the mature athlete, can be improved only slightly. Therefore, in order to increase speed, the runner must increase stride length.

Stride length may be increased in two ways:

1. Overstriding—putting the leading foot ahead of the body's center of mass on each stride—does increase stride length, but it also reduces stride frequency drastically. The result is *always* a slower total time for the race.

2. Pushing against the ground with greater force (and, of course, the ground pushing back equally) increases stride length without necessarily decreasing stride frequency. The

Figure 48. To get maximum force, the shot putter's feet should not leave the ground until the shot has left his hand. The putter on the left is getting maximum force. The putter on the right is losing a great amount of force which must instead be absorbed by the body.

athlete who can push against the ground harder with each stride increases his stride length naturally and runs faster. It is possible, even for the mature athlete, to reduce his 100-yard time through heavy leg strength exercises. The coach who says "sprinters are born, not made" has not considered leg strength and stride length.

When stride frequency and stride length are considered, it is interesting to note that world class sprint races are always decided on stride length—not frequency. Slow motion analysis of close stretch runs shows that all good sprinters slow down about equally (see page 20) and that they all take about the same number of strides in a given amount of time. It is the sprinter who takes the *fewest strides in a given distance* who wins the race. Obviously, then, muscular strength is a more important factor in sprinting than is muscular endurance.

Another interesting consideration is the stride length of the hurdler. The distance between hurdles determines the length of his stride, no matter how much stronger his legs become. Therefore, one might conclude "it is easier for a sprinter to improve than it is for a hurdler." But, what we are not considering is that it is much easier for the hurdler to increase stride frequency than it is for the sprinter, because there are ten unusually unorthodox strides in his race. In order to improve his speed, he *must* improve stride frequency. And this he does by reducing the time it takes to clear the hurdles, which is, in effect, an increase in stride frequency.

IMPULSE

It is not only the amount of force applied to the ground (or an implement) that determines speed, but also the *time* during which that force is applied. (Force x Time = Impulse.)

In the jumping and throwing events and in sprint starting, the best techniques are those in which the greatest possible force is applied for the longest possible time.

The development of the O'Brien style of shot putting, for example, allowed the putter to exert muscular force over a greater distance (and, thus, for a longer time) than was possible with previous styles. The result was greater impulse and longer puts.

In the straddle high jump take-off, a bent lead leg delivers greater force to the ground than a straight lead leg, but the time of that force is very brief. The force delivered by a straight lead leg, even though it is not as great, is maintained for a longer period of time. The result is greater impulse and higher jumps.

CHAPTER 15

Centrifugal Force

Centrifugal force is the outward pulling force on a rapidly turning object (or person). It is the force that holds water in a bucket swung rapidly over the head in a science class experiment.

In track and field coaching, the classic example of centrifugal force is the hammer thrower's struggle to hold on to the hammer. The faster the hammer thrower turns, the "heavier" the hammer becomes and the more difficult it is to control. Ultimately, though, it is the centrifugal force which helps to determine how far the hammer can be thrown. (See Figure 49.)

In the running events, running is easiest (and fastest) when the running surface is perpendicular to the force applied to it. This creates a centrifugal force problem when a sprinter is running an unbanked curve, however, since the necessary inward lean of the athlete (compensating for centrifugal force) applies the force at an angle to the running surface. This explains why runners are never able to run quite as fast on curves as on straightaways. (See Figure 50.)

The Fosbury flop curve approach takes advantage of centrifugal force. The flopper follows a curved path on his very fast approach run. When he plants his foot and leaves the ground, his body continues in a line tangent to his approach curve. The centrifugal force, coupled with checking linear motion (See Chapter 9), puts him in his lay-out position over the crossbar.

Figure 49. The hammer thrower takes advantage of centrifugal force.

Figure 50. For one to run as fast on curves as on straightaways, the curves must be banked enough so that the force of the leaning runner (compensating for centrifugal force) is perpendicular to the banked surface. The sharper the curve, the greater the bank must be. On a flat track, the runner in the inside lane, because he is running a sharper curve, must lean more than those in the outside lanes. Therefore, the runners in outside lanes have an advantage.

Glossary

ACCELERATION is positive rate of change of velocity (speeding up).

AERODYNAMICS is the branch of physics which treats the laws of motion of bodies in air.

An **AXIS** is a straight line about which a body rotates.

An **AXIS OF MOMENTUM** is the axis of a rotating body in the air, passing through the center of mass, about which all parts rotate in one direction until contact with the ground is regained.

An **AXIS OF MOVEMENT** is the axis of a body in the air, passing through the center of mass, around which a movement originating in the air will bring about an equal and opposite movement on the opposite side of the center of mass.

CENTER OF MASS is the single point in a body where all mass on any side of the body is equal to the mass on the opposite side. (Also called *CENTER OF GRAVITY*.)

CENTRIFUGAL FORCE is the force pulling outward during rotation.

CHECKING LINEAR MOTION is the process through which rotary momentum may be begun when linear motion is interrupted at one end of a moving body and the other end continues at an increased rate of speed.

COMPONENT VELOCITIES are an upward and a forward ground velocity which, when added together, determine angle and velocity through the air.

CONSERVATION OF ROTARY MOMENTUM is a law which states that a rotating body has a constant rotary momentum—the product of rotary inertia and rotary velocity.

DECELERATION is negative rate of change of velocity (slowing down).

DYNAMICS is the branch of physics which deals with the action of force on bodies. (Often called *MECHANICS,* especially in Europe.)

ECCENTRIC THRUST is off-center thrust.

EFFECTIVE FORCE is the force actually "received" by the ground or implement.

FORCE (Mass times Acceleration) is any physical cause which modifies the motion of a body.

The **FRONTAL AXIS** is the axis which passes through the body's center of mass from side to side.

GROUND REACTION is the equal and opposite reaction of the ground to any force against the ground.

HORIZONTAL VELOCITY is velocity in a horizontal direction (forward).

IMPULSE is Force times Time.

INERTIA is the resistance to change in motion.

An **INERTIAL AXIS** is an axis which has great rotary inertia, so that its reaction will be through fewer degrees than the original action.

LINEAR MOTION is straight line movement from one point to another.

The **LONGITUDINAL AXIS** is the axis which passes through the body's center of mass from head to toe.

MASS is the amount of matter in an object.

MOMENTUM is Mass times Velocity.

MOTION is continuous change of position.

NEWTON'S 1ST LAW—(The Law of Inertia) A body at rest or in uniform straight-line motion will continue in that state until compelled to change by an external force.

NEWTON'S 2ND LAW—(The Law of Acceleration) Acceleration is proportional to force and inversely proportional to mass.

NEWTON'S 3RD LAW—(The Law of Reaction) For every force there is an equal and opposite force of reaction.

NUTATION is rotation in the air around two axes at the same time (conical fashion), caused by an eccentric thrust.

An **OPTIMUM ANGLE** is the angle or release (or jumping) which provides the maximum possible desired height or distance.

A **PARABOLIC CURVE** is the regular curve followed by an object's center of mass when projected into the air.

PRIMARY AXES are axes around which the entire body must rotate when free in the air. The three primary axes, which are mutually perpendicular, are the longitudinal axis, the frontal axis and the saggital axis.

A **RESULTANT VELOCITY** is the sum of two component velocities.

ROTARY INERTIA is the resistance to change in rotary motion. (*MOMENT OF INERTIA* in physics.)

ROTARY MOMENTUM is Rotary Inertia times Rotary Velocity. (*ANGULAR MOMENTUM* in physics.)

ROTARY MOTION is the turning movement of an object around an axis located within the object itself. (*ANGULAR MOTION* in physics.)

ROTARY VELOCITY is the angle through which a body turns in one second (revolutions per second). (*ANGULAR VELOCITY* in physics.)

The **SAGGITAL AXIS** is the axis which passes through the body's center of mass from front to back.

A **SECONDARY AXIS** is an additional axis created by the body at a distance from the center of mass.

TRANSFERENCE is the process through which rotary momentum may be transferred, usually from one part of the body to the entire body.

VERTICAL VELOCITY is velocity in a vertical direction (upward).

Bibliography

BADE, EDWIN, *The Mechanics of Sport,* Kingswood, England: Andrew George Elliot, 1952.
BUNN, JOHN W., *Scientific Principles of Coaching,* Englewood Cliffs, N.J.: Prentice-Hall, Inc., 1955.
BROER, MARION, *Efficiency of Human Movement,* Philadelphia: W.B. Saunders Co., 1966.
DYSON, GEOFFREY, *The Mechanics of Athletics,* London: University of London Press, 1968.
GAYNOR. FRANK, *Concise Dictionary of Science,* Totowa, New Jersey: Littlefield, Adams & Co., 1967.
GRISWOLD, LARRY, *Trampoline Tumbling,* St. Louis: Fred Medart Co., 1948.
HOPPER, BERNARD J., *The Dynamical Basis of Physical Movement,* Twickenham, England: St. Mary's College Physics Laboratory, 1959.
HORNE, DENNIS E., *Trampolining,* London: Faber & Faber, Ltd., 1968.
JENSEN, CLAYNE R., AND SCHULTZ, GORDON W., *Applied Kinesiology,* New York: McGraw-Hill Book Co., 1970.
LADUE, FRANK, AND NORMAN, JIM, *Two Seconds of Freedom.* Brentwood, England: Nissen Trampoline Co.. Ltd.. 1967.
MCCLOY, C.H., *Mechanical Analysis of Physical Education Activities,* unpublished and undated.
RASCH, PHILIP J., AND BURKER, ROGER K., *Kinesiology and Applied Anatomy,* Philadelphia: Lea and Febiger, 1967.
ROGERS, ERIC M., *Physics for the Inquiring Mind,* Princeton, N.J.: Princeton University Press, 1960.
TRICKER, R.A.R., AND TRICKER, B.J.K., *The Science of Movement,* London: Mills and Boon, Ltd., 1967.
WARTENWEILER, J., JOKL, E., AND MEBBELINCK, M., *Bio-Mechanics,* Basel, Switzerland: S. Karger, 1968.
WILT, FRED, *Mechanics Without Tears,* Tucson, Arizona: United States Track and Field Federation, 1970.
WILT, FRED, AND ECKER, TOM, *International Track and Field Coaching Encyclopedia,* West Nyack, N.Y.: Parker Publishing Co., 1970.

Index

Acceleration, 19-22: and forward lean, 21-22; negative, 20; of falling bodies, 86; of sprinter, 20; positive, 20

Action-reaction in the air, 72-79: flop high jumping, 43-45, 77-78; long jumping, 88; pole vaulting, 42-43; straddle high jumping, 72-73, 75-76

Action-reaction on the ground, 96-102: high jumping, 96-98; shot putting, 98, 100; sprinting, 96-97; sprint starting, 97

Aerodynamic forces: air resistance, 21-24; disci and javelins, 36-37, 38, 87

Air resistance, 21-24: aerodynamic forces, 36-37, 38; in sky diving, 83

Angle: discus and javelin, 36-37; hammer throwing, 36; optimum release, 36-38; optimum take-off in long jump, 37-38; release, 26-27; shot putting, 36-37; take-off, 24

Axes, 39-45: flop high jumping, 43-45; inertial, 80-85; nutation, 45; of momentum, 42; of movement, 42; pole vaulting, 42; primary, 39, 41; secondary, 91-95; straddle high jumping, 42-43

Backward lean, 22-24
Baseball: eccentric thrust, 62

Center of gravity (See Center of mass)
Center of mass, 28-31: "flight path," 32; high jumping, 30-31; human body, 28-31; human body, 28-31; long jumping, 30; parabolic curve, 32-36; shot, 28

Centrifugal force, 103-104: curve running, 103-104; flop high jumping, 103-104; hammer throwing, 103-104

Checking linear motion, 64-68: flop high jumping, 65-66; forward rotation, 64-68; javelin throwing, 65; long jumping, 89-90

Component velocities, 24-27, 36: high jumping, 24; hurdling, 24-25; long jumping, 24; sprint starting, 24

Conservation of rotary momentum, 51-60: discus throwing, 55, 57; divers and trampolinists, 52, 57, 59-60; flop high jumping, 57; hang style long jumping, 57; on turntable, 57-58; pirouetting ice skater, 51, 52, 56; pole vaulting, 52, 54, 57; sprinting, 52, 54; straddle high jumping, 52-53

Curve running, 103-104
Curves of flight, 32-38: aerodynamic forces, 38; hurdling, 33, 35-36; optimum angles, 36-38; parabolic curve, 32-36

Deceleration, 19-22: of sprinter, 21, 101
Discus: aerodynamic forces, 36-37, 38; angle of release, 36-37; eccentric thrust, 64; hollow, 50; nutation, 45; parabolic curve, 36; rotary inertia, 50; rotary motion, 18
Discus throwing: conservation of rotary momentum, 55, 57; re-

109

lease angles, 36-37; transference of momentum, 69, 71
Diving: conservation of rotary momentum, 52, 59; eccentric thrust, 62-63; inertial axes, 81, 83
Dynamics: definition, 9-10, 15; importance of, 17
Dyson, Geoffrey H. G., 7, 10, 17

Eccentric thrust, 61-64: baseball, 62; demonstrating, 63-64; discus, 64; diver, 62-63; flop high jumping, 66-67; football, 64; long jumping, 67-68; straddle high jumping, 62-63, 66-67
Ecker, Dr. Richard E., 10

Falling bodies, 86-90: acceleration of, 86; long jumping, 86-90; pole vaulting, 83; sky diver, 87
Football: eccentric thrust and wobble, 64; momentum in, 47-48
Forward lean, 21-24
Forward rotation: checking linear motion, 64-68; long jumping, 88, 93-94
Fosbury, Dick, 9, 67, 68, 103

Galileo, 10, 86
Ground reaction, 96-102: high jumping, 96-102; shot putting, 98, 100; sprinting, 96-97; sprint starting, 97

Hammer: parabolic curve, 36
Hammer throwing: angle of release, 36; centrifugal force, 103, 104
Hang style long jumping, 57, 95
High jumping (general): component velocities, 24; ground reaction, 96; lay-out, 30-31, 34; low center of mass, 30-31; parabolic curve, 33-35; take-off, 24, 96, 97-98; western roll, 30-31
High jumping (flop), 9, 67, 68, 103: action-reaction, 43-45, 77-78; axes, 43-45; axis of momentum and movement, 43-45; checking linear motion, 65-66; centrifugal force, 103-104; conservation of rotary momentum, 52-53, 57; curve approach, 103-104; eccentric thrust, 66-67; lay-out, 65-67, 69; take-off, 65-67, 69-70, 77-78; transference of momentum, 69-70
High jumping (straddle): action-reaction in the air, 72-73, 75-76; axes, 42-43; axis of momentum, 42; conservation of rotary momentum, 52-53; eccentric thrust, 62-63, 66-67; ground reaction, 96-98; impulse, 102; lay-out, 30-31, 45, 52-53, 61-62, 73, 75-76; nutation, 45; rotary momentum, 52; rotary motion, 18; take-off, 61-62, 69, 97, 102; transference of momentum, 69
Hitch-kick, 93-95
Hopper, Bernard, 7, 10
Horne, Dennis, 17
Human body, center of mass in, 28-31
Hurdling: action-reaction, 78-79; component velocities, 24-25; form, 83-85; inertial axes, 83-85; lowering head, 35-36; parabolic curve, 33, 35-36; stride length, 101

Ice skater, 51, 52, 56
Impulse, 101-102: jumping, 101; shot putting, 101-102; sprint

starting, 101; straddle high jumping, 102
Inertia, 49-50: in sprinting, 49; of discus, 50; of shot, 50; rotary, 49-50
Inertial axes, 80-85: astronaut, 80; cat or rabbit, 80; diver, 83; hurdling, 83-85; rabbit, 81-83; sky diving, 83; trampolinist, 80

Javelin: aerodynamic forces, 36-37, 38, 87; nutation, 45; parabolic curve, 36
Javelin throwing: angle of release, 36-37; checking linear motion, 65
Jumping: impulse, 101; linear and rotary movements in, 19

Lean: backward, 22-24; forward, 21-24, 91, 92
Linear motion, 18
Long jumping: action-reaction, 88; center of mass, 30; checking linear motion, 89-90; component velocities, 24; eccentric thrust, 67-68; forward rotation, 88-90; hang style, 57, 95; hitch-kick, 93-94; holding the legs up, 86-88; optimum take-off angle, 37-38; parabolic curve, 33, 34, 35; secondary axes, 91-93; somersault style, 67-68; take-off, 37-38; transference of momentum, 69; use of arms, 35, 94

McCloy, Dr. C. H., 10
Momentum, 46-48: in football, 47-48; in sprint starting, 46-47; in weight men, 47; rotary, in high jumping, 52
Momentum, axes of, 42: in flop high jumping, 43-45

Motion, 18-27: linear, 18; rotary, 18
Movement, axis of, 42: in flop high jumping, 43-45; in pole vaulting, 42

Newton, Sir Isaac, 10, 72, 80, 96
Nutation, 45

Optimum angles, 36-38: long jumping, 37-38

Parabolic curve (or parabola), 32-36: high jumping, 33-35; hurdling, 33, 35-36; long jumping, 33, 34, 35; of disci, 36; of hammers, 36; of javelins, 36; of shots, 36
Pirouetting ice skater, 51, 52, 56
Pole vaulting: action-reaction, 42-43; axis of movement, 42; conservation of rotary momentum, 52, 54, 57; Dooley's pole bending, 9; speed of fall, 83
Primary axes, 39, 41

Rabbit, falling, 80, 81-83
Release angles, 26-27, 36-38: discus and javelin throwing, 36-37; hammer throwing, 36; optimum, 36-38; shot putting, 26-27, 36-37
Resultant velocity, 24-27
Rotary motion, 18: in high jumping, 18; of discus, 18
Rotary inertia, 49: of discus, 50
Rotary momentum in high jumping, 52
Rotation, forward: checking linear motion, 64-68; long jumping, 88, 93-94
Running: linear and rotary motion in, 18; on curves, 103-104

Secondary axes, 91-95: long jumping, 91, 93; roller skater, 91; ski jumper, 91-92; sprinting, 91, 92; triple jumping, 94-95; wire walker, 91
Shot: application of force to, 27; center of mass, 28; inertia, 50; parabolic curve, 36
Shot putting: action-reaction on the ground, 98, 100; component velocities, 26-27; impulse, 101-102; O'Brien style, 9, 101; release angle, 26-27, 36-37
Skating, 51, 52, 56
Ski jumping, balance in, 91-92
Sky diving: air resistance, 83; falling speed, 87; inertial axes, 83
Somersault long jumping, 67-68
Sprinting: acceleration, 20; action-reaction on the ground, 96-97; conservation of rotary momentum, 52, 54; deceleration, 21, 101; increasing speed, 99-101; inertia in, 49; linear motion in, 18; secondary axes, 91, 92; stride frequency, 99; stride length, 99; wind-aided, 24
Sprint starting: action-reaction on the ground, 97; component velocities, 24; impulse, 101; momentum, 46-47
Stride length: hurdling, 101; sprinting, 99
Stride frequency: in sprinting, 99

Take-off: angle, 24; high jump, 24, 96, 97-98; (flop, 65-67, 69-70, 77-78); (straddle, 61-62, 69, 97, 102); long jump, 37-38
Throwing: impulse, 101; linear and rotary motion in, 19; weight man's mass, 47
Trampolining: conservation of rotary momentum, 52, 57, 59-60; inertial axes, 80-81
Transference of momentum, 68-71: discus throwing, 69, 71; flop high jumping, 69-70; long jumping, 69; straddle high jumping, 69
Triple jumping: hitch-kick, 94-95; secondary axes, 94-95
Turns from the ground, 69-71
Turns in the air, 72-79: action-reaction, 72-79; flop high jumping, 77-78; hurdling, 78-79; straddle high jumping, 72-76
Turntable: action-reaction demonstration, 72, 74; conservation of rotary momentum demonstration, 57-58

Velocities: component, 24-27, release, 26-27; resultant, 24-27

Western roll, 30-31
Wilt, Fred, 10
Wind-aided races, 24
Wind resistance (See air resistance)